The Webb Space Telescope

A New Era in Cosmic Exploration and Discovery

Sam Holly

Copyright

About the Author

Sam Holly is a well-known figure in both science and literature, where he has made important accomplishments. Sam has a background in astrophysics and has spent his whole career trying to figure out the mysteries of the universe. His easy-to-understand writing style often helps people understand difficult science ideas.

Sam started his academic career with a degree in physics. He then went on to get a Ph.D. in astronomy, which is the study of cosmic events and how they affect our understanding of space and time. His study has helped us understand things like dark matter, how galaxies form, and the lifecycle of stars, which is a big part of what we know about the subject.

Sam has a knack for making the complicated world of the universe easy to understand in a way that both interests and teaches his readers. In many of his books, he writes about the areas where science, philosophy, and history meet. These areas show how astronomical findings have affected human culture and thought over time.

Sam is also in high demand as a teacher and speaker. He is known for giving interesting lectures and public talks that make people appreciate the wonders of the world. He supports teaching science and reading and writing because he thinks that an informed public is important for the growth of humanity.

Sam's contributions to science and literature have not only advanced our knowledge of the cosmos but have also encouraged a deeper appreciation for the beauty and complexity of the universe we inhabit. His writing and research continue to influence scientists and writers of today and tomorrow. It embodies the spirit of exploration and curiosity that moves human inquiry forward.

Table of Content

Introduction to the Webb Space Telescope

Overview of the Webb Space Telescope

The James Webb Space Telescope (JWST), often referred to simply as the Webb Space kTelescope, marks a monumental leap forward in our ability to observe the universe. As the most powerful space telescope ever built, Webb is intended to peer into the very origins of the universe, capturing light from the first galaxies that formed after the Big Bang. This remarkable instrument, equipped with a 6.5-meter primary mirror made of 18 hexagonal segments of gold-coated beryllium, can view objects too old, distant, or faint for the Hubble Space Telescope.

Webb operates mainly in the infrared spectrum, allowing it to see through cosmic dust and gas clouds that obscure many celestial bodies, revealing the hidden structures of the universe. Its suite of sophisticated instruments and sensors allows scientists to study the atmospheres of exoplanets, possibly identifying signs of habitability, and to observe the formation of stars and galaxies in unprecedented detail. Positioned at the second Lagrange point (L2), approximately 1.5 million kilometers from Earth, Webb benefits from a stable thermal climate and an unobstructed view of space.

Historical Context and Development

The conception of the Webb Space Telescope goes back to the late 20th century, following the launch of the Hubble Space Telescope. Recognizing Hubble's limitations and the need for a more advanced observational platform, the astronomical community began to imagine a next-generation space telescope that could view the universe in the infrared spectrum. This vision was driven by the desire to look further back in time, to the universe's infancy, and needed technology that could operate at extremely cold temperatures to be sensitive to infrared light.

The development of the Webb Space Telescope involved extensive international collaboration, led by NASA with major contributions from the European Space Agency (ESA) and the Canadian Space Agency (CSA). This ambitious project faced numerous challenges, including technical hurdles linked to its innovative design, the complexity of its deployment mechanism, and the need to work at cryogenic temperatures. After decades of research and rigorous testing, the Webb Space Telescope was launched on December 25, 2021, aboard an Ariane 5 rocket from French Guiana. Its successful deployment and commissioning marked the beginning of a new era in astronomy, with the promise of revolutionizing our understanding of the universe's earliest days.

Chapter 1

Design and Technology of the Webb Space Telescope

The James Webb Space Telescope (JWST), often referred to as Webb, stands as a pinnacle of astronomical engineering, pushing the boundaries of what humanity can view in the cosmos. Its design and technology are a testament to decades of innovation, collaboration, and ingenuity, aimed at answering some of the most profound questions about our world.

Advanced Optics and Mirrors

At the heart of Webb's observational ability is its advanced optical system, centered around a primary mirror that spans 6.5 meters in diameter. Unlike traditional telescopes that use a single-piece primary mirror, Webb's mirror is formed of 18 hexagonal segments made from beryllium, chosen for its remarkable stability under extreme cold conditions. Each segment is coated with a thin layer of gold to optimize the reflection of infrared light, increasing the telescope's sensitivity to the faintest signals from the faraway universe.

This segmented design allows the telescope to achieve a large collecting area necessary for observing the earliest stars and galaxies while keeping the ability to be compactly folded within the confines of a rocket fairing for launch. After reaching its target in space, Webb's mirror segments unfold and align with nanometer accuracy, a process controlled by a sophisticated actuation system. This alignment is important for the telescope to produce sharp and accurate images.

The secondary mirror, another important component of Webb's optical system, further focuses the light collected by the primary mirror onto the telescope's instruments. The exact curvature and positioning of this mirror are important for the optimal performance of the observatory, ensuring that the light from the edges and center of the primary mirror converges at the same point in the instrument section.

Cutting-Edge Instruments and Sensors

Webb is equipped with a suite of four main scientific instruments, each designed to make a wide range of observations across the infrared spectrum:

1 Near-Infrared Camera (NIRCam):

NIRCam serves as Webb's main imaging instrument, capable of capturing high-resolution images in the near-infrared range. It is equipped with a range of filters to isolate specific wavelengths of light, allowing astronomers to identify the physical and chemical properties of celestial objects.

2 **Near-Infrared Spectrograph (NIRSpec):** NIRSpec can view up to 100 objects simultaneously, dispersing their light into spectra. This capability allows the detailed study of the chemical composition, temperature, and motion of stars and galaxies, giving insights into their physical conditions and evolution.

3 **Mid-Infrared Instrument (MIRI):** MIRI combines both imaging and spectroscopy capabilities in the mid-infrared range, where it can identify features that are inaccessible to NIRCam and NIRSpec. MIRI's sensitivity to longer wavelengths allows it to peer through dust clouds that obscure many regions of star formation and planetary systems, showing phenomena hidden from view in the near-infrared.

4 Fine Guidance Sensor/Near InfraRed Imager and Slitless Spectrograph (FGS/NIRISS):

This instrument is dual-purpose, built both to stabilize the telescope's line of sight and to perform wide-field imaging and slitless spectroscopy. The FGS ensures Webb's exact pointing capability, while NIRISS offers unique observing modes for studying exoplanet atmospheres and faraway galaxy clusters.

Each of these instruments includes an array of sensors and detectors optimized for infrared observations, allowing Webb to capture data across a broad spectrum of wavelengths. These detectors are kept at extremely low temperatures to minimize noise and maximize sensitivity, allowing

Webb to identify the faintest whispers of light from the edge of the observable universe.

Thermal Shield and Sun Protection

Webb's ability to view the universe in infrared is critically dependent on its thermal environment. To keep its instruments at the necessary cryogenic temperatures, Webb is equipped with a massive sunshield the size of a tennis court. This five-layered shield is made from a specially coated material that reflects sunlight, keeping the telescope's viewing side in perpetual shadow and at temperatures as low as 40 Kelvin (-233°C, -388°F).

The sunshield's design is not only a marvel of heat engineering but also of mechanical complexity. This process was carefully tested on the ground, but its successful operation in the harsh environment of space is a testament to the ingenuity and precision of Webb's design.The combination of Webb's advanced optics, cutting-edge instruments, and innovative thermal protection systems allows it to conduct observations that were previously beyond reach. By capturing the light from the first galaxies, studying the formation of stars and planetary systems, and probing the atmospheres of exoplanets, Webb is not only expanding our knowledge of the universe but also paving the way for future generations of space telescopes.

Chapter 2

Mission Objectives of the Webb Space Telescope

The James Webb Space Telescope (JWST), with its unprecedented observational capabilities, is poised to transform our knowledge of the universe. Its mission goals are ambitious, focusing on unraveling cosmic riddles that have puzzled astronomers for decades.

Exploring Distant Galaxies

One of Webb's main objectives is to explore the most distant galaxies in the universe, providing a glimpse into the era shortly after the Big Bang. These early galaxies are key to understanding how the cosmos grew over time. Webb's advanced infrared sensitivity allows it to identify the faint light from these ancient celestial bodies, whose light has been stretched into the infrared spectrum by the expanding universe.

By observing these distant galaxies, Webb aims to answer critical questions about their formation, evolution, and the part they play in the cosmic web. It will study their shapes, sizes, colors, and ages, giving insights into the processes that drove the evolution of the universe from a smooth state to the complex structure we observe today. This exploration will not only shed light on galaxy formation but will also help astronomers understand the distribution of dark matter and the impact of dark energy on the universe's expansion.

Studying the Formation of Stars and Planets

Another important objective of the Webb Space Telescope is to study the formation of stars and planetary systems. Webb's infrared powers allow it to peer through dense clouds of dust and gas where stars are born. By observing these stellar nurseries in unprecedented detail, Webb will show the initial conditions and processes that lead to star formation.

Beyond individual stars, Webb is also focused on finding the secrets of planet formation. By observing protoplanetary disks—rings of gas and dust surrounding young stars—Webb can identify the signatures of planet formation, such as gaps and rings in the disks where planets may be coalescing. This research will enhance our knowledge of the diversity of planetary systems and the conditions that lead to the formation of habitable worlds, possibly shedding light on our own Solar System's origins.

Investigating the Origins of the Universe

Webb's mission stretches to investigating the very origins of the universe, including the conditions that prevailed soon after the Big Bang. By observing the cosmic microwave background radiation and the most distant observable galaxies, Webb will provide insights into the universe's initial state and its later evolution.

This objective includes not only looking at the earliest light in the universe but also studying the fundamental physics that governs cosmic structures. Webb will study how the first stars and galaxies influenced the reionization of the universe, ending the cosmic dark ages and making the universe transparent to light. This line of research will help astronomers understand the timeline of the universe's key events, from the Big Bang to the present day.

In addition to these main objectives, Webb's mission is expected to uncover new phenomena and cosmic objects that have yet to be imagined. Its observations will likely raise new questions, furthering the search of knowledge about the universe. The Webb Space Telescope, through its exploration of distant galaxies, study of star and planet formation, and investigation of the universe's origins, stands as a beacon of human curiosity and ingenuity, poised to open the secrets of the cosmos.

Chapter 3

Significant Discoveries by the Webb Space Telescope

The James Webb Space Telescope (JWST), with its unparalleled observational capabilities, is set to revolutionize our knowledge of the universe. Its mission, meant to peer deeper into space and further back in time than ever before, is expected to yield a trove of significant discoveries.

Unveiling New Star-Forming Regions

One of the early expected achievements of Webb is the unveiling of previously unseen star-forming regions within our own galaxy and beyond. These regions, often enshrouded in thick dust clouds, have stayed largely hidden from optical telescopes. Webb's advanced infrared imaging and spectroscopy capabilities allow it to penetrate these dust clouds, showing the intricate processes of star birth.

By observing these regions, Webb will provide vital insights into the initial conditions and physical processes that lead to star formation. This includes the collapse of gas and dust into protostars, the accretion of material onto these forming stars, and the ultimate ignition of nuclear fusion. Additionally, Webb's observations will enhance our knowledge of how massive stars influence their surroundings, including the effect of their intense radiation and powerful outflows on the formation of subsequent stellar generations.

Observing Exoplanetary Atmospheres

Another area where Webb is expected to make major strides is in the observation of exoplanetary atmospheres. With its sensitive instruments, Webb can study the light from stars as it passes through the atmospheres of transiting exoplanets. This analysis, known as transmission spectroscopy, shows the chemical composition, temperature, and pressure profiles of these distant worlds' atmospheres.

Webb's observations are set to detect and characterize a wide range of atmospheric constituents, including water vapor, carbon dioxide, methane, and perhaps even signs of more complex organic molecules. These discoveries are important for understanding the conditions on these planets and assessing their possible habitability. Moreover, Webb's detailed exoplanet studies will add to our wider understanding of planetary system formation and evolution, offering comparative insights with our own Solar System.

Deep Field Observations and Findings

Building on the legacy of the Hubble Space Telescope's deep field observations, Webb is expected to push the limits of deep space imaging. By observing select patches of the sky for extended periods, Webb will collect light from the most distant galaxies, giving a window into the universe's infancy.

These deep field observations will likely show galaxies that formed just a few hundred million years after the Big Bang, giving critical data on their sizes, shapes, masses, and formation rates. Such results will help astronomers piece together the evolutionary timeline of the cosmos, shedding light on the era of reionization, the role of dark matter in galaxy formation, and the assembly of large-scale cosmic structures.

The significance of these findings cannot be overstated. Each will add a vital piece to the puzzle of our universe's story, from the smallest scales of planetary atmospheres to the vast tapestry of cosmic evolution. As Webb continues its mission, the scientific community eagerly expects a new era of astronomical breakthroughs, each promising to expand our knowledge of the cosmos and our place within it.

Chapter 4

Impact of the Webb Space Telescope on Modern Astronomy

The James Webb Space Telescope (JWST), often hailed as the top observatory of the next decade, is set to redefine the contours of modern astronomy. Its unprecedented capabilities in observing the universe in the infrared spectrum place it as a pivotal instrument in advancing astrophysical theories, enriching cosmology, and deepening our knowledge of dark matter and dark energy. This transformative effect is anticipated across various domains of space science, promising to unlock answers to long-standing questions and to ask new ones in the relentless quest to comprehend the cosmos.

Advancements in Astrophysical Theories

The JWST is expected to provide vital insights that will refine and, in some cases, revolutionize existing astrophysical theories. By observing the universe's earliest galaxies, stars, and planetary systems, the telescope will offer empirical data that could either support or challenge current models of cosmic evolution. For instance, the detailed images and spectra of the earliest galaxies will test ideas about galaxy formation and evolution, possibly altering our understanding of how complex structures emerged from the primordial universe.

Moreover, Webb's observations of star-forming regions and protoplanetary disks will improve our knowledge of the processes that lead to star and planet formation. This includes the dynamics of accretion disks, the role of magnetic fields in star formation, and the myriad pathways that lead to the varied range of planetary systems discovered by astronomers. By giving a clearer picture of these processes, the JWST will allow scientists to refine theoretical models, leading to a more nuanced understanding of the physical laws that govern the universe.

Contributions to Cosmology

Cosmology, the study of the universe's origin, structure, evolution, and eventual fate, stands to gain immensely from the

JWST's observations. The telescope's deep field imaging potential, akin to peering back in time, will allow astronomers to study the universe's conditions shortly after the Big Bang. This could yield unprecedented insights into the inflationary period of the universe, a brief epoch of rapid expansion that cosmological models predict happened fractions of a second after the Big Bang.

Additionally, the JWST's detailed observations of the Cosmic Microwave Background (CMB) radiation—the afterglow of the Big Bang—and its ability to identify the earliest light-emitting objects will provide valuable data to test theories about the universe's early stages. This includes the nature of cosmic reionization, a crucial era when the first stars and galaxies ionized the intergalactic medium, ending the cosmic dark ages.

Enhancing Our Understanding of Dark Matter and Dark Energy

Perhaps two of the most enigmatic components of the universe, dark matter and dark energy together comprise about 95% of the total cosmic content, yet remain poorly known. The JWST is set to shed light on these dark constituents through its observations of their gravitational effects on visible matter and the universe's expansion rate.

By mapping the distribution of galaxies and watching the bending of light around massive galaxy clusters—a phenomenon known as gravitational lensing—Webb will provide clues about the nature and distribution of dark matter. These observations could help refine models of large-scale structure formation in the universe, giving insights into the role of dark matter in binding galaxies together.

Regarding dark energy, the force thought to be responsible for the accelerated expansion of the universe, the JWST will study distant supernovae and the large-scale structure of the cosmos. These observations will help astronomers measure the rate of cosmic expansion more accurately, testing theories of dark energy and its effects on the fate of the universe.

Broader Implications

The impact of the JWST goes beyond the enhancement of specific theories and models. Its comprehensive survey of the cosmos will produce vast datasets, allowing a data-driven approach to astronomy that will complement theoretical models with empirical evidence. This synergy between observation and theory is expected to accelerate the pace of finding, leading to a more dynamic and nuanced area of astronomy.

Furthermore, the JWST's mission will likely inspire a new breed of telescopes and observational platforms, both in space and on the ground. The technological innovations created for the JWST, from its sunshield to its segmented mirror design, will serve as a blueprint for future observatories aiming to study the universe in even greater detail.

The James Webb Space Telescope marks a monumental leap forward in our quest to understand the cosmos. Its effect on modern astronomy will be felt across multiple domains, from the refinement of astrophysical theories to groundbreaking contributions to cosmology and the study of dark matter and dark energy. As the JWST begins to unveil the universe's secrets, it will surely redefine our place within the cosmic tapestry, ushering in a new era of discovery and wonder in the field of astronomy.

Chapter 5

Challenges and Solutions in the James Webb Space Telescope Mission

The journey of the James Webb Space Telescope (JWST) from idea to reality is a story rich with challenges and triumphs, underscoring humanity's relentless pursuit of knowledge and our capacity for innovation. The JWST, meant to be the most sophisticated space telescope ever built, faced a myriad of scientific and operational challenges throughout its development. These challenges required ingenious solutions and adaptations, showcasing the ingenuity and resilience of the teams involved.

Technical and Operational Challenges

1 Complex Deployment Mechanism:

The JWST's design includes a complex deployment mechanism, necessary for unfolding its big sunshield and mirror array. This mechanism had no parallels in previous space missions, requiring the invention of new technologies and extensive testing to ensure reliability in the vacuum of space, far from any chance of manual intervention.

2 Thermal Management:

Operating mainly in the infrared spectrum, the JWST must be kept at extremely low temperatures to detect faint celestial signals without interference from the telescope's own thermal emissions. Designing a thermal management system that could maintain these cryogenic temperatures in the harsh environment of space presented a major engineering challenge.

3 Vibration and Launch Stresses:

The telescope's intricate instruments and delicate mirror segments had to survive the intense vibrations and mechanical stresses of launch. Ensuring the stability of these sensitive components, especially the segmented primary mirror's alignment, was crucial for the mission's success.

4 Long-Distance Communications:

Positioned at the second Lagrange point (L2), approximately 1.5 million kilometers from Earth, the JWST needs robust and reliable communication systems.

Maintaining consistent contact and data transmission over such vast distances, especially with the high volumes of scientific data expected, offered a sizable challenge.

Innovative Solutions and Adaptations

1 Deployment Rehearsals:

To handle the complexities of the JWST's deployment mechanism, engineers performed exhaustive rehearsals on the ground. These simulations involved full-scale models and were meant to mimic the zero-gravity environment of space. Every step of the deployment process was carefully planned, practiced, and refined to minimize the risk of failure.

2 Sunshield Design:

The answer to the JWST's thermal management challenge came in the form of a five-layer sunshield, each layer as thin as a human hair. This sunshield, roughly the size of a tennis court, reflects the Sun's heat away from the telescope, keeping the instruments at cryogenic temperatures. The materials and design of the sunshield were groundbreaking, mixing thermal protection with the need for minimal weight.

3 Vibration Isolation and Testing:

To protect the JWST's sensitive components from launch-induced vibrations, engineers created advanced vibration isolation systems. Additionally, the telescope underwent rigorous shake and acoustic tests that mimicked the launch conditions. These tests ensured that every component could survive the journey to space without compromising the telescope's delicate optics and sensors.

4 High-Gain Antenna and Data Management:

To solve the challenges of long-distance communications, the JWST is equipped with a high-gain antenna capable of transmitting large volumes of data back to Earth. The mission's operations center uses sophisticated data management tools to handle, process, and distribute the scientific data collected by the telescope. This infrastructure ensures that the vast amounts of information gathered by the JWST are efficiently relayed to scientists around the world for study.

Broader Impact of Solutions

The innovative solutions created for the JWST have broader implications beyond the mission itself. The technologies and methodologies pioneered during the telescope's development are adding to advancements in various fields, from aerospace engineering to data management. The experience gained from designing and deploying the JWST's complex systems will inform future space missions, possibly reducing risks and costs connected with deploying large, intricate structures in space.

In conclusion, the James Webb Space Telescope's trip from conception to launch is a testament to human ingenuity in the face of daunting challenges. The technical and operational hurdles encountered during the mission's development led to creative solutions and adaptations that not only ensured the telescope's successful deployment but also advanced our capabilities for future space exploration. As the JWST starts its mission to unravel the mysteries of the universe, it stands as a symbol of what humanity can achieve when curiosity, creativity, and collaboration meet.

Chapter 6

Future Prospects of the James Webb Space Telescope

The James Webb Space Telescope (JWST) represents a monumental leap in our capability to observe the universe, but its journey is just starting. The telescope's deployment and early operations have set the stage for a future rich with scientific discovery and cooperation. Looking ahead, the JWST's effect is expected to stretch far beyond its immediate findings, influencing upcoming missions and fostering collaboration with other space observatories.

Upcoming Missions and Extended Goals

The success of the JWST paves the way for a new age of space missions, each building on the technological advancements and scientific discoveries of its predecessor. Future telescopes and space observatories will likely aim to complement and build upon the JWST's capabilities, focusing on different wavelengths, regions of space, or specific cosmic phenomena. For example, missions might target the high-energy universe (X-rays and gamma rays) more widely or focus on the detection and detailed study of exoplanets using direct imaging methods.

One of the extended goals of the JWST itself is to continue pushing the boundaries of what we can view. As the telescope's mission progresses, it may be guided to explore more deeply into the origins of the universe, the nature of dark matter and dark energy, and the potential for life on planets orbiting other stars. The telescope's design includes a degree of flexibility, allowing for the possibility of new observing modes and strategies to be created in reaction to scientific discoveries and community needs.

Additionally, there's the possibility for extending the JWST's operational life beyond its initial 10-year goal. Through careful management of fuel and resources, and assuming the telescope's instruments and systems continue to work optimally, the

JWST could continue to provide invaluable data for many years to come.

Collaboration with Other Space Observatories

The JWST is not a solitary observer of the world but rather a key player in a network of telescopes and observatories, each with its unique strengths. Collaborative observations with facilities like the Hubble Space Telescope, the Chandra X-ray Observatory, and the upcoming Extremely Large Telescopes (ELTs) on the ground will provide a more complete understanding of astronomical objects and phenomena.

One area of collaboration could involve simultaneous or sequential observations of particular targets, combining the JWST's infrared powers with the optical, ultraviolet, or X-ray observations of other telescopes. This multi-wavelength approach can show different aspects of cosmic objects and events, from the structure of galaxies and the dynamics of star formation to the composition of exoplanet atmospheres.

Furthermore, the data from the JWST will complement that of other projects focused on cosmology and fundamental physics, such as the Euclid mission and the Laser Interferometer Space Antenna (LISA). By working in concert, these observatories can handle some of the most pressing questions in astrophysics, such as the nature of dark matter and dark energy, the properties of black holes, and the distribution of galaxies in the cosmic web.

Envisioning the Future

The future prospects of the James Webb Space Telescope encompass a broad spectrum of possibilities, from extending its own mission limits to playing a crucial role in a network of space- and ground-based observatories. As the JWST starts to fulfill its mission objectives, it will surely inspire future missions designed to explore unanswered questions that arise from its discoveries. The telescope's legacy will be measured not only by the scientific breakthroughs it achieves but also by the way it shapes the future of space travel and our knowledge of the universe. The JWST's journey is a beacon for future efforts in space science, heralding an age of unprecedented exploration and discovery in the decades to come.

Conclusion

The James Webb Space Telescope (JWST) stands as a monumental achievement in the field of space exploration and observation, marking a new chapter in humanity's desire to understand the cosmos. Its sophisticated design, groundbreaking capabilities, and the wealth of knowledge it promises to deliver will leave an indelible mark on the legacy of space science and pave the way for future efforts in astronomy.

The Legacy of the Webb Space Telescope

The legacy of the JWST will be defined by its unparalleled contributions to our knowledge of the universe. By peering back in time to observe the light from the first galaxies, studying the formation of stars and planetary systems, and probing the atmospheres of distant exoplanets, Webb is set to answer some of the most basic questions about the nature of the cosmos and our place within it. These contributions will not only improve our understanding of how the universe works but will also inspire a new generation of scientists, engineers, and ordinary people to look up at the stars with renewed wonder and curiosity.

Moreover, the technological innovations created for the JWST, from its sunshield and segmented mirror to its intricate deployment mechanisms, mark important advancements in aerospace engineering. These technologies have potential applications beyond astronomy, including in the areas of telecommunications, materials science, and even climate monitoring. The JWST's mission exemplifies how pushing the boundaries of space exploration can drive technological innovation and add to advancements across multiple fields.

The Future of Space Exploration and Observation

The future of space travel and observation is bright, with the JWST laying the groundwork for the next generation of space telescopes and missions. Its successes and challenges will guide the design and operation of future observatories, both in orbit and on the ground. As we look ahead, we can expect missions that will build on the JWST's discoveries, exploring new wavelengths, employing even more advanced technologies, and perhaps even venturing to new vantage points within our solar system and beyond.

The collaborative nature of the JWST's mission—encompassing contributions from NASA, the European Space Agency, the Canadian Space Agency, and numerous other international partners—highlights the importance of global collaboration in the pursuit of scientific knowledge. This collaborative approach is likely to be a defining feature of future space missions, as the complexity and cost of cutting-edge space exploration increasingly necessitate joint efforts by countries and agencies around the world.

In conclusion, the James Webb Space Telescope is more than just an observatory; it is a beacon of human ingenuity and a testament to our combined desire to discover the unknown. Its legacy will be measured not only in the scientific findings it makes but in the way it inspires future generations to continue the quest for knowledge. As we stand on the brink of a new era in space exploration, the JWST tells us of the vastness of the universe awaiting our gaze and the endless possibilities that lie within our reach.